Connecting the Bible and Mathematics through Data Analysis (9 – 12)

Connecting the Bible and Mathematics through Data Analysis (9 – 12)

Javier L. Taylor

Black Jack Publishing
Atkins, Arkansas

Editorial, Sales and Customer Service office

Black Jack Publishing
2284 Worthen Road
Atkins, AR 72823
blackjackpress@gmail.com

Find us on Facebook!

All Scripture quotations and references are from the King James Version of the Bible.

ISBN 10: 0615428983
ISBN13: 978-0615428987

How does mathematics relate to the Bible?

While the Bible is not intended as a math textbook, it is rich its use of numbers, shapes, and three dimensional structures. We need to realize that people have been using sophisticated mathematics since the beginning. God created us with a natural curiosity and desire to count, design, and build. We have the opportunity to explore how mathematics might have been used thousands of years ago.

What is data analysis?

In his book "*Introduction to Data Analysis*" Joel Levine describes data analysis as "a body of methods that help to describe facts, detect patterns, develop explanations, and test hypotheses." Data analysis can be thought of as consisting of two parts: exploration and inference (Cobb 1997). Data analysis is attempting to make sense of data. To collect, classify, summarize, and present this data we will often use statistical measures. Using these measures allows us to compare, make conjectures and even predictions about the nature of the data.

Why is data analysis important?

We are bombarded with bits of data every day through television, ads in the newspaper and magazines, or even reports we read for work. Daily we look at charts, tables of values, the claims of ads on television, in magazines, or in the newspaper. How do we wade through the data in the world around us? The study of data is worthwhile simply because of the amount of data we encounter in everyday life. Data analysis introduces us to statistical tools we can use to understand the data we are being fed. We will consider statistical tools such as Mean, Median, Mode, Mean Absolute Deviation, Standard Deviation, Spread, Range, and Interquartile Range to make sense of data.

What is my purpose in putting these activities together?

The purpose of this book is to use the Bible as a context for discussing several of the statistical tools available to a high school student. I believe it is vitally important that we use the Bible as a context to discuss spiritual things, like holy living, like our service to God, etc. However, that said, t it is also vital that our children see the Bible as relevant in other areas of life. I think this is what God was telling Moses to relate to Israel in Deuteronomy 6: 6 – 8. We are to discuss the Bible in many contexts including math and science. When you look at the data concerning church kids (children who attended church at least twice a week through high school) when they go off to college, it is alarming to say the least. We must battle for the future of our children by presenting them with the Word of God in a way that answers their questions and challenges them to examine the world around them carefully in light of the Holy Scripture. I hope this book gives the Christian school teacher or the Christian homeschooler activities which can be used to present the Bible as the Living Word of God.

Why worry about integrating the Bible into the study of mathematics?

Biblical integration is probably one of the most difficult tasks for a secondary mathematics teacher. When I was in the classroom I felt like my time was restricted and that many of the topics I needed to cover were not directly covered in the Scripture. I needed to cover topics like factoring quadratics, solving logarithmic equations, or determining the derivative of an equation. I would often use Scripture readings in class so that I kept the Word of God in the class, but I always felt like I didn't connect the Scripture to the mathematics. Disconnecting the subject matter from the Scripture forces the students to compartmentalize their faith, something like "the Bible is okay in Bible class but it has no relevance in math class."

When I read the 2010 Barna Group Research results, I was heartbroken. The result of compartmentalizing our faith is that many Christians today are scripturally illiterate. We have to face the fact that when someone challenges the Scripture they are challenging our faith. We have to teach our children to have an answer for those who ask about or even challenge their faith (1 Peter 3: 15). A Christian's faith and the Word of God cannot be separated and the Christian remain whole. The Christian has to trust the Bible in every single area. Jesus, himself, said "the scripture cannot be broken" (John 10: 35).

Where does the study of data analysis fit into a mathematics curriculum?

The study of data analysis in grades 9 – 12 should build on student work through grade 8. Students should already be familiar with summary statistics, data graphs such as line plots, bar graphs, scatter plots, histograms, box plots, and frequency charts.
High school data analysis should focus attention on statistical reasoning. Students should be expected to generalize patterns and draw conclusions based on exploratory skills developed in earlier math courses. The project-based activities in this book assume students can identify and describe measures of central tendency, variation, and spread.
The National Council of Teachers of Mathematics (NCTM) identifies principles that serve to guide and foster the use of the activities in "*Connecting the Bible and Mathematics Through Data Analysis (9 – 12).*"
Students should:

- Understand the need for and be able to formulate well-defined questions.
- Study the role that randomization plays in producing data
- Identify assumptions
- Have hands-on experiences in producing data
- Be involved in organizing and analyzing data

The activities in ""'*Connecting the Bible and Mathematics Through Data Analysis*" are designed to help students select and use appropriate statistical tools to analyze data, develop and evaluate inferences and to predictions that are based on data.

About the activities

- **Numbering the Days of a Man, Pt. 1**

This activity is based on Genesis 5: 1 – 32 and allows students an opportunity to examine a data set that consists of the ages of the pre-flood patriarchs. Students should calculate the central tendencies of the data set as well as plot the data as a scatter plot, histogram and box plot. The students should be encouraged to have discussions about the different representations of the data set and what each plot shows and hides. The students should determine the measures of spread, variation, and deviation and discuss the implications of the variance.

- **Numbering the Days of a Man, Pt. 2**

This activity is based on Genesis 11: 10 – 32 and is a continuation of Number the Days of a Man, Part 1part 1 and allows students an opportunity to find measures of central tendency, spread, variation, and deviation with another set of data as well as make some conjectures about the distribution of the data. Students are encouraged to compare the data sets from Genesis chapter 5 and Genesis chapter 11.

- **Estimating Pi**

This activity is based on 1 Kings 7: 23 – 26 and allows the students to use make measurements and use statistical tools to evaluate the variation inherent in multiple measures. The students will then use this information to challenge the misconception that God says Pi is three by considering the passage concerning the Molten Sea. The students will determine a reasonable estimate of Pi from the data they collected.

- **A Study of the Cubit and Span**

This activity is based on Genesis 6: 16 and Exodus 28: 18 ands focuses on the development of the students understanding of the distribution of a data set. Students should discuss what it means to vary from the "norm" and the importance of setting a standard unit of measure to allow for conformity in measurement. The students should also discuss the differences in the Mean Absolute Deviation and the Standard Deviation.

- **Modeling the Pre-flood Population**

This activity is based on Genesis 1: 27 – 28 and allows students the opportunity to make conjectures and models of early population growth, discuss the various constraints, make inferences and defend their conjectures. This should allow students to argue for a particular model based on their constraints for the problem.

- **Calculating Usury**

This activity is based on Leviticus 25: 356 – 37 and Ezekiel 18: 8 – 17 and gives students the opportunity to see how debt grows and consider debt in light of Scripture and in light of society. Students should discuss interest especially predatory interest. The students should discuss why God might have a problem with this kind of lending practice.

- **The Power of Sharing the Gospel**

This activity is based on Matthew 28: 19 and allows students the opportunity to see the power of sharing the gospel with other people and how the gospel message can travel. The spread investigated here is a geometric spread like a virus. The students should see that when we follow the Great Commission given in Matthew 28: 19 we can change the world very quickly.

- **Calculating Israel's Population Growth in Egypt**

This activity is based on Genesis 46: 27, Exodus 1: 5 – 9, and Numbers 1: 46 and allows students the opportunity to make and test conjectures about the possible population growth rate of Israel while they were living in Egypt.

Mathematical Goal:
To guide students to consider mean, median, mode, variation and deviation of a data set.

Student Learning Expectations:
- Students will make a Histogram of the ages of the patriarchs found in Genesis 5: 1 - 32
- Students will make a box and whisker plot of the ages of the patriarchs found in Genesis 5: 1 – 21
- Students will determine the mean absolute deviation and the standard deviation of the ages
- Students will estimate the mean lifespan of a pre-flood patriarch

Topic question:
What was the pre-flood lifespan?

Cooperative Groups: 3-4 students per group	**Materials Needed:** Centimeter grid Paper Chart paper Pencil Graphing Calculator
Questions to Consider: How will students represent (show) their thinking? How will students communicate (tell) their thinking? How will students reason (explain) their thinking?	**Vocabulary:** Minimum, Maximum, Range Mean, median, mode Variation Deviation Mean absolute deviation Standard deviation Outlier

Lesson Timeline:

Time	Activity	Notes
2-3 min	Do you know about how long a person lives today in the US? Do you think it was always this way?	The mean lifespan of a person is available on the Internet at www.cdc.gov The students might have grandchildren who have lived a long time so some discussion about how long a person can live is appropriate.
5 min	The students should discuss what an average lifespan means what are things that affect our lifespan? What do you think was the average lifespan of someone in the Old Testament? Was it different before the flood? If it was, why do you think it was different?	It is difficult to conceive of a person living for over 900 years, but we can see in Genesis that this was the case. Use the histogram to find if there are clusters in the data. You can also use the life of Enoch to discuss outliers and their effect on the average and the median. A good question might be: If we looked at lifespan data from today would Methuselah be an outlier? Explain.

15 min	Group the students in groups of 3 – 4 and have them create a histogram (you could have them do a Stem and Leaf chart instead). Either should show the data cluster(s). Allow students to use chart paper to record their work. If students also do a stem and leaf on chart paper, it can be turned sideways as a lead-in to histograms	Letting students get in groups after writing on their own, allows students to share ideas, and get help if they are struggling with the writing.
20 min	**Explore part 1:** Have the students find the median of the whole data set. Then have them find Quartile 1, the median between the min and the median of the whole data set. Then have them find Quartile 3, the median between the median of the whole data set and the max of the data set. These three points, along with the maximum and minimum values, form the five number summary used to build a box and whisker plot. **Explore part 2:** Have the students find the mean of the data set and then build a table showing the difference between the mean and each of the data points.	Possible Questions: How are the individual ages different from the average age? How can we find a number that summarizes this difference? One way is to find the mean of the absolute value of the square of each difference (see table provided). What are advantages and disadvantages of an absolute value or a difference squared?
15 min	Summarize: Each group will share their displays. What can be easily seen from in the displays? How easy are they to read? Do they have titles? Do they have labels? Can they make true data statements from the displays?	Bring out vocabulary when appropriate.

Next Day Extension:
What do we do about Enoch? In this data set Enoch can be a problem especially for the mean of our data set. What happens to the mean absolute deviation, standard deviation, mean, median and mode when Enoch is removed from the data set? What if Methuselah were removed?

Use the ages data from Genesis to estimate Enoch's lifespan if he had not been taken by God. Explain why your estimate is reasonable using the data or a display.

What was the lifespan of the pre-flood patriarchs?

1. Read Genesis 5: 1 – 32 and use the table provided to record the ages of the patriarchs found in the chapter. Construct a scatter plot of the data set.

2. Describe the clusters of the data you see in the scatter plot. How are the ages distributed? Find the median of the set and circle that number. Find the median from the minimum age up to but not including the median you found before and circle that number. This is called Quartile 1 (Q1). Find the median from the median of the whole data set (the first one you find but not including that number) and the maximum of the set is called Quartile 3 (Q3). You now have three medians and the minimum and the maximum. Use these to construct a box-and –whisker plot. Find the Inter Quartile Range of the set (Q3 – Q1).

3. Using the table provided find the standard deviation and the mean average deviation. The mean average deviation is found by minding the mean of the absolute value of the differences, Standard deviation is found by finding the square of each difference, finding the mean of differences and then taking the square root of that mean. Why do you think these numbers are different?

4. What is your estimate of the lifespan of a pre-flood patriarch? Support your answer with the graphs and calculations you made in earlier problems. How accurate is your value? Explain.

5. How quickly do you find the ages dropping? Why is this do you think?

Name	Age	$(\text{Age} - \text{Mean age})^2$	$\lvert \text{Age} - \text{Mean Age} \rvert$
Adam			
Seth			
Enos			
Cainan			
Mahalleel			
Jared			
Enoch			
Methuselah			
Lamech			
Noah			

Mathematical Goal:
To guide students to think about the use of mean, median, mode, range, variance, mean absolute deviation, and standard deviation

Student Learning Expectations:
- Create a scatter plot, histogram, and box and whisker plot of the age data in Genesis 11: 1 - 32
- Find the Interquartile range of the Genesis 6 age data and the Genesis 11 data
- Choose a graph in comparing the age data found in Genesis 5 with that found in Genesis chapter 11.

Topic question:
What was the mean age of a person before the flood?

Cooperative Groups: 3-4 students per group	**Materials Needed:** Centimeter grid paper Pencil Graphing Calculator Sticky pads Chart paper
Questions to Consider: How will students represent (show) their thinking? How will students communicate (tell) their thinking? How will students reason (explain) their thinking?	**Vocabulary:** Variation Interquartile range Data distribution Histogram Box and Whisker Plot Scatter plot

Lesson Timeline:

Time	Activity	Notes
10 min	If people in Bible days lived such long lives, why are our life spans so much shorter? When did life spans begin to decrease? Show data and a Scripture reference to support your conclusion.	Discuss the decline in ages found in Genesis chapter 6 and the decline found in chapter 11 which data set is decreasing faster?
5 min	Each student will determine how to best compare the data sets	An example of this type of comparison is: "Which 10th grade class is the tallest?" What do we need to know before we can answer this question
15 min	Place students in groups of 3 or 4 and have them share their plan with each other, then with the whole group. Also have them discuss their data collection plan and make decisions about what to do next.	The reason for grouping after they have written on their own is for pairing up students who might be stuck with the writing with others who have ideas.
20 min	Explore Part 1. Collect data Explore Part 2. Represent data using a graph of their choice. Bar graph, frequency table (line plot)…	Students will go to a station with paper and pencil. They will take turns recording and measuring.
15+ min	Summarize: Each group will share their graphs. What does the graph tell us? How easy are they to read? Do they have titles? Do they have labels? Can they make true data statements from the graph?	Bring out vocabulary when appropriate.

Data Analysis and Statistical Modeling
Numbering the Days of a Man, Pt. 2 (Genesis 11: 10 – 32)

Next Day Extension:
In groups have the students search the Internet or some other place for two data sets they want to compare i.e. the height of players of two basketball teams, …etc. Display their findings on chart paper and present to the class.

What was the lifespan of the Post-flood patriarchs?

1. Read Genesis 11: 10 – 32 and make a table of the ages of the patriarchs found in the chapter. Use this table to construct one of the following: Scatter plot, and a Box and Whisker plot

2. Describe the clusters of the data you see. Is there a graph that allows you to see the clusters more clearly? How are the ages distributed? Find the median of the set and circle that number. Find the Inter Quartile Range of the set.

3. Using the table provided find the standard deviation and the mean average deviation. Why do you think these numbers are different?

4. What is your estimate of the lifespan of a post-flood patriarch? Support your answer with the graphs and calculations you made in earlier problems. How accurate do you believe your value is? Explain.

5. How quickly do you find the ages dropping? Why do you think this is?

Name	Age	$(\text{Age} - \text{Mean age})^2$	$\lvert \text{Age} - \text{Mean Age} \rvert$
Shem			
Arphaxad			
Salah			
Eber			
Peleg			
Reu			
Serug			
Nahor			
Terah			
Abraham			

Mathematical Goal:
To guide students to find an estimate of Pi through statistical modeling

Student Learning Expectations:

- Students will measure the length from their elbow to the tip of their fingers in centimeters (to the nearest millimeter)
- Students will find their hand span in centimeters (to the nearest millimeter)
- Students will analyze the classroom data and chose a length for their classroom cubit based on the classroom data (mean, median, mode, etc)

Topic question:
Does the Bible imply that Pi is 3?

Cooperative Groups: 3-4 students per group	**Materials Needed:** Centimeter grid Paper Chart paper Pencil Graphing Calculator
Questions to Consider: How will students represent (show) their thinking? How will students communicate (tell) their thinking? How will students reason (explain) their thinking?	**Vocabulary:** Mean, median, mode variation deviation mean absolute deviation Standard deviation Outlier Proportionality Ratio

Lesson Timeline:

Time	Activity	Notes
2-3 min	Discuss the passage. Some critics of the Bible claim this passage implies that the value of Pi is 3. Is this true? Can we trust everything the Bible says?	If we can trust that God created the universe and made a way for us to be saved then can we can trust all the Bible
10 min	The students need to take all the measurements and record them. Measurements need to be made very carefully and accurately.	The cubit was the measure of a man's arm from the elbow to the tip of the fingers. The hand span was a measure from the tip of the thumb to the tip of the pinky finger while the hand is fully extended
15 min	Group the students in groups of 3 – 4 and have them create a Stem and Leaf, histogram, or a Box and Whisker plot to should show the data cluster(s). Use chart paper to have the students record their work	Putting the students in groups allows the students to express and defend their ideas, as well as the giving them the opportunity to share their thinking about the constraints of the problem.
20 min	**Explore part 1:** Have the students use the data to determine their cubit size. It is okay if different groups use a different measure for the length of a cubit. **Explore part 2:** Have the students read the passage carefully and note the width of the Molten Sea and	The students should be able to defend their choice of cubit based on the collected data. Don't give them a "right" answer because there isn't one the data will have to drive their

	whether it is reasonable to measure brim to brim from the inner brim or the outer brim. Determine the way to measure the circumference was this the inner circumference or the outer circumference? Make several models of the measurements and calculate the values of Pi.	choice.
15+ min	Summarize: Each group will share out their graphs. What does the graph tell us? How easy are they to read? Do they have titles? Do they have labels? Can they make true data statements from the graph?	Bring out vocabulary when appropriate

Next Day Extension:
Different ancient cultures had different lengths for a cubit. The smallest length I have found is 17 inches the largesl is about 22 inchcs. How does the length of the cubit compare with that of the hand span? Do you think there is a proportional relationship? Are there other body measurements that are proportional? Why do you think this is?

Does God know about Pi?

1. Some critics of the Bible have claimed that God implies that Pi is 3. Read 1 Kings 7: 23 – 26. Verse 26 tells us that the sea was a "hand breadth thick" which must be accounted for when calculating the diameter and circumference.

2. With a meter stick find the length of everyone's forearm from the elbow to the tip of the forefingers in centimeters. Then use the meter stick to find the measure of everybody's hand breadth. To measure this extend your hand fully and measure from the tip of the thumb to the tip of the pinky finger.

3. Once you have everybody's measurements create a table and a graph of the data to find the clusters of the data and choose one measurement of a cubit and handbreadth to represent the group's data.

4. If the measurements of the sea were made from inner brim to inner brim what would be the value for the ratio of the circumference and the diameter? Use your representative measurements to see what the ratio of circumference and diameter would be if you built the sea.

Data Analysis and Statistical Modeling
Estimating Pi (1 Kings 7: 23 – 26)

Mathematical Goal:
To guide students to make population models based on constraints.

Student Learning Expectations:
- Students will identify issues affecting population growth in the pre-flood world
- Students will choose a population model and identify the limiting factors of the model
- Students will build a population model for the pre-flood world

Topic question:
How did the pre-flood population grow?

Cooperative Groups: 3-4 students per group	**Materials Needed:** Paper Pencil Graphing Calculator
Questions to Consider: How will students represent (show) their thinking? How will students communicate (tell) their thinking? How will students reason (explain) their thinking?	**Vocabulary:** Mathematical Modeling Constraints Resolution of models

Lesson Timeline:

Time	Activity	Notes
2-3 min	Do you know the population of the city you live in? What about the population of the United States? How could you find out if you didn't know?	Since we have just gone through a census in 2010 it should not be difficult to have population figures available on the Internet
10 min	The students should discuss one method for finding the population of a city or state. What difficulties may arise? The students should individually write a plan for determining the population of their state.	Some of the constraints are that we cannot be everywhere at once to count the people. We also have to account for the fact that people are constantly being born, while others are dying, so the population is fluid from second to next. The resolution of a model is similar to the resolution of a photo it is difficult to see the details of a low resolution photo but easy to see detail in a high resolution photo. Models also have a resolution; the resolution is based on the information available and the constraints of the problem.
15 min	Group the students in groups of 3 – 4 and have them determine a plan for finding the population of the earth just before the flood in Genesis chapter 6.	Putting the students in groups allows the students to express and defend their ideas, as well as the giving them the opportunity to share their thinking about the constraints of the problem.
20 min	Explore part 1: Starting with a population of two people, what would the population be in 100 years? What are some of the assumptions we must make in modeling this population?	

	How does God's judgment in Genesis 3: 16 affect the population growth? What are some recent events that might have affected the population growth? Before the fall, man would have lived forever. Do you think God reduced the birthrate at that time? Why would this be necessary? Have the students make a chart or a graph showing their population growth from Adam and Eve until Noah and his family. Are the students' models reasonable?
15+ min	Summarize: Each group will share out their graphs. What does the graph tell us? How easy are they to read? Do they have titles? Do they have labels? Can true data statements be made from the graph?

Next Day Extension:
Compare the plans of students at different grade levels. How do they use the mathematics they know to model this problem? Does a more sophisticated model necessarily mean a better model?

If Adam and Eve each lived for over 900 years and they were able to have children for only half that time (an assumption) and they had a child every five years (another assumption) how many children would they have had? Use this same assumption for all the patriarchs of Genesis 3 – 6 and their offspring, what would the population of the earth be at the time of the flood of Genesis 6? Do you think this model will predict a population that is higher or lower than the actual population? Explain your reasoning.

The Bible records three of Adam and Eve's children (Cain, Abel, and Seth) does this mean they only had three children? Have a Bible verse to defend your reasoning

Have you ever thought about how populations grow?

1. Using the website http://2010.census.gov/2010census/data/ find the population growth for your city and state for the time period of 2000 to 2010.

2. There are always constraints that effect population in any given area. For people it might be things like opportunities for employment, crime rate, opportunities for education, etc. When you consider constraints of the population in the Garden of Eden, what do you think of?

3. In Genesis 3: 16, God told Eve that He would greatly multiply her sorrow and her **conception.** Why do you think God would keep the conception rate low while Adam and Eve lived in the Garden?

4. Why would God increase the conception (birth rate) once Adam and Eve were removed from the Garden?

5. What was the key thing that changed in the lives of Adam and Eve when they sinned?

6. Today women are able to have children for roughly half of their lives. Since people lived much longer in the days of Adam and Eve how long do you think they were able to have children?

7. If Eve lived about as long as Adam, and was able to have children for half of her life, how many years could she have had children?

8. If Adam and Eve had a child every 5 years how many children would they have had?

Mathematical Goal:
To guide students to analyze the cubit and span and guide the students to compare the measurements of the students cubits and spans and the class mean cubit and span

Student Learning Expectations:

- Students will measure the distance from their elbow to the tip of their fingers and the measure the distance from the tip of their pinky finger to the tip of their thumb with their hand fully extended.
- Students will make a table of the different cubit measures of the class members and a table of the different span measurements of the class members.
- Students will analyze the spread of the cubit and span measurements
- Students will find the class mean cubit and span and compare their individual measurements to the class mean.
- The students will determine the variance of the measures of cubit and span and quantify the variance.
- The students will determine and compare the standard deviation and the mean absolute deviation of the measurements of cubit and span.

Topic question:

What is a cubit anyway?

Cooperative Groups: 3-4 students per group	**Materials Needed:** Centimeter grid Paper Chart paper Pencil Graphing Calculator Meter Stick
Questions to Consider: How will students represent (show) their thinking? How will students communicate (tell) their thinking? How will students reason (explain) their thinking?	**Vocabulary:** Gaussian Distribution Interquartile Range Skew

Lesson Timeline:

Time	Activity	Notes
2-3 min	Remind students about the cubit and demonstrate the process of finding the measurement of the cubit and span	Discuss how a cubit could be used as a linear measure and the span
5 min	The students make their measurements and record them. Begin a class table so the students can make their comparisons	You will also want to measure the height of each student for the extension problem
10 min	Group the students in groups of 3 – 4 and have them determine the mean cubit for the class and then to make a table of the differences between their individual measurements and the class mean	Putting the students in groups allows the students to express and defend their ideas, as well as the giving them the opportunity to share their thinking about the constraints of the problem.
30 min	**Explore part 1:** Have the students build a box-and-whisker plot of the data and determine the spread of the cubit and span data sets **Explore part 2:** Have the students use the differences to find the variance of the data set. Use the table included to help the students find the Standard deviation and the Mean absolute deviation of the cubit and span data sets. Have the students make a scatter plot of the cubit versus span data is there a relationship between the length of a person's cubit and the length of their span?	How does the spread of the data compare with the variance of the data sets? How does the standard deviation differ from the mean absolute deviation?
15+ min	Summarize: Each group will share their models and discuss the accuracy of the models as well as the process they used to model the data. Discuss the strengths and weakness of the respective models.	Bring out vocabulary when appropriate

Next Day Extension:

Have the students determine if a relationship exists between the length of a person's cubit and their height. Have the students build a mathematical model to predict height of a person based on the length of their cubit. How accurate is the model? Have the students decide if a similar model can be constructed using the span data.

Just what is a cubit anyway?

1. What is the spread of the cubit data set for the class?

2. Separate the cubit data into girls' measurements and boys' measurements, which set has greater spread?

3. Separate the span data into girls' data and boys' data. Calculate the mean absolute deviation for the span data for each set. Which set shows the greatest deviation?

4. How does the spread of the span data compare with the variance of the data? Explain.

5. How does the standard deviation of the cubit data compare with the mean absolute deviation? Explain.

Data Analysis and Statistical Modeling

A Study of the Cubit and Span (Genesis 6: 16; Exodus 28: 18)

Name	cubit Measure	(Cubit – class mean)^2	$\lvert$Cubit – Class Mean$\rvert$

Data Analysis and Statistical Modeling

A Study of the Cubit and Span (Genesis 6: 16; Exodus 28: 18)

Name	Span Measure	(span – class mean)^2	$\lvert$ Span – Class mean $\rvert$

Name	Cubit Measure	Difference from Class Mean	Span Measurement	Difference from Class Mean

Mathematical Goal:
To guide students to think about how the interest affects the amount of a loan to be repaid.

Student Learning Expectations:

- Students will develop the standard interest formula $A = P(1 + r)^t$
- Students will find an interest rate that causes a loan to double in 10 years
- Students will determine the interest made from a loan of given principle and time
- Students will determine the original loan amount given the payback time and interest rate

Topic question:
Why do you think the Law forbade excessive Usury among the Israelites?

Cooperative Groups: 3-4 students per group	**Materials Needed:** Centimeter grid Paper Chart paper Pencil Graphing Calculator
Questions to Consider: How will students represent (show) their thinking? How will students communicate (tell) their thinking? How will students reason (explain) their thinking?	**Vocabulary:** Principal Interest rate Usury

Lesson Timeline:

Time	Activity	Notes
2-3 min	Do you know the average interest rate for a credit card?	Find several credit card rates and use them in this discussion
5 min	The students will discuss the use of credit cards and indebtedness in view of the following Scriptures: Leviticus 25: 36 – 37; Ezekiel 18: 8, 13, 17	How does God feel about a Christian being in excessive debt?
15 min	Group the students in groups of 3 – 4 and have them develop the standard interest (annual) from the problem " I borrowed $100 dollars at 6% interest if I make no payment how much do I owe in 1 year, 2 years, etc. (see attached table and graph)	Putting the students in groups allows the students to express and defend their ideas, as well as the giving them the opportunity to share their thinking about the constraints of the problem.
20 min	**Explore part 1:** Have the students calculate the interest necessary for a loan payback amount to be double in 10 years. Assume the original amount to be $10,000. Can a general statement be made about the interest necessary to double a loan in 10 years? (Working with the interest formula the students determined in the group activity) **Explore part 2:** Have the students calculate the interest made from a $1,000 loan paid off over 2 years at 8% interest. (compound interest) **Explore part 3:** If the payback for a four year loan at 6% is 2040.73 what the amount of the money borrowed? What would have been the payback of the loan if the interest rate had been 8.5%? (simple interest)	

15+ min	Summarize: Each group will share out Their solutions to the explore problems. What do they notice about how interest affects the payback amount? Why do you think God has a problem with high interest?	Bring out vocabulary when appropriate

Next Day Extension:
Have the students determine the amount of money they would have they invested $3,500 at 3.5% interest when they are 18 years old and keep the money in savings until they are 65 years old.

Why do you think the Law forbade excessive Usury?

1. What do you think about credit card use? Do you know the current interest rate on most credit cards? Write a paragraph which states your belief about the role of credit in the life of a Christian.

2. Find the amount owed after 2 years if \$1,000 is borrowed at 6% interest compounded monthly. Use the table provided to help you with these calculations.

3. Determine an interest rate that will cause the payback amount to be double the original loan amount in 10 years (annual payments). Calculate the same for double payback in 5 years.

4. Determine the original loan amount if the loan is repaid in 3 years with 6.5% interest and the payback amount is \$3,568.51.

1000(1+.06)					

Explain why this expression works using the distributive property as part of your explanation.

Mathematical Goal:
To guide students to think about an exponential growth model and to compare exponential growth models to linear growth models.

Student Learning Expectations:

- Students will graph an exponential equation
- Students will explore a doubling model with a table of values
- Students will determine how many in their city could be reached if everyone in their school witnessed to two people

Topic question:
How can we reach the world with the Gospel?

Cooperative Groups: 3-4 students per group	**Materials Needed:** Centimeter grid Paper Chart paper Pencil Graphing Calculator
Questions to Consider: How will students represent (show) their thinking? How will students communicate (tell) their thinking? How will students reason (explain) their thinking?	**Vocabulary:** Exponential growth Exponential graph $Y = ab^x$ Linear Growth $Y = mx + b$

Lesson Timeline:

Time	Activity	Notes
2-3 min	How do you think God intended for us to reach the world with the Gospel?	Ask for students to describe our responsibility to share the gospel. Have students find Scriptures to support this.
5 min	Discuss the possibility of reaching just two people with the gospel and teaching them to each reach two people teaching these two to reach two, and so on. How do you think the graph of time and the number of people reached would look?	Ask the students to express what the graph might look like and keep a record of these.
15 min	Group the students in groups of 3 – 4 and have them use the provided table to develop the standard exponential equation $y = ab^x$.	The reason for grouping after they have written on their own is for pairing up students who might be stuck with the writing with others who have ideas.
20 min	**Explore part 1:** Have the students calculate how many days the gospel would need to be shared to reach their entire city? Their state? And The United States of America **Explore part 2:** If a group of 12 people shared with two friends and trained them to share with two friends and so on, how many days to reach everyone in the world?	This does make the assumption everyone who hears the gospel gets saved. The class could discuss whether this is reasonable or not. When modeling with Math sometimes these assumptions are made discuss this along with the population modeling and the

		underlying assumptions
15+ min	Summarize: Each group will share out their graphs. What does the graph tell us? How easy are they to read? Do they have titles? Do they have labels? Can they make true data statements from the graph?	Bring out vocabulary when appropriate.

Next Day Extension:

What would happen if you shared with two people each day, but two did not share with others? Compare the graph and equation of this model with the earlier one. How important is it that sharing the Gospel doesn't stop with you?

What do you think was God's plan for us to reach the world with the gospel?

1. Imagine that you witness to just two people and you teach them to each witness to two people. What do you think the graph or table of this situation would look like if the witnessing happened daily?

2. Using the table provided determine a mathematical formula for the number of people witnessed to in "x" days.

3. How many days would it take for the Gospel to reach a number of people equal to the population of your city? Your state?

4. If twelve people determined to share the gospel with two people each day and teach them to each share with two other people per day, how many days would it take for the gospel to reach a number of people equal to the population of the world?

Complete the Chart

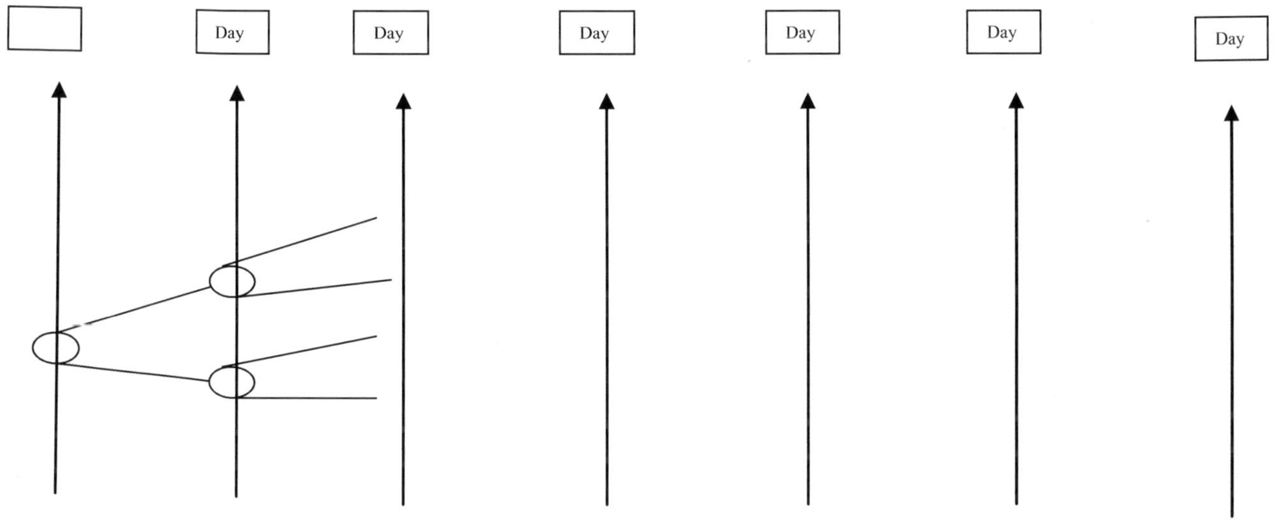

General Equation for an exponential Function:_______________

Mathematical Goal:
To guide students to purposefully model a population growth

Student Learning Expectations:

- Students will make predictions about the population growth based on the information found in the Bible
- Students will determine the mean family size of the children of Jacob
- Students will use the mean family size to set the boundaries for the population growth of Israel

Topic question:
What was the population growth rate is Israel while in Egypt?

Cooperative Groups: 3-4 students per group	**Materials Needed:** Centimeter grid Paper Chart paper Pencil Graphing Calculator
Questions to Consider: How will students represent (show) their thinking? How will students communicate (tell) their thinking? How will students reason (explain) their thinking?	**Vocabulary:** Mean, median, mode Variation Deviation Mean absolute deviation Standard deviation Outlier

Lesson Timeline:

Time	Activity	Notes
2-3 min	What is the population growth rate for your city according the last census?	The population data is available on the Internet at www.census.gov
5 min	The students should discuss the passages in Genesis 46 and in Numbers 1 and Exodus 1: 1 – 9. Each student should write a paragraph about what population size would make Pharaoh so nervous.	How fast would the population have to have grown for the king of Egypt to become so concerned about the size of Israel?
15 min	Group the students in groups of 3 – 4 and have them determine the population growth rate for the different tribes and the growth rate for the entire nation. Which tribe has the largest growth rate?	The reason for grouping after they have written on their own is for pairing up students who might be stuck with the writing with others who have ideas.
20 min	**Explore part 1:** Have the students find the mean family size of the children of Jacob. Then have the students use the mean to build a population model and determine when they think Pharaoh might have become nervous. **Explore part 2:** Have the students estimate the population of the nation of Israel when they marched out of Egypt. Have the students make a table of the population in 10 year intervals.	Questions: how fast was the population growing for Israel while in Egypt? What growth rate do you get when you use 430 years starting population of 70 and a mean family size of 4 children? 5 children? 6 children?
15+ min	Summarize: Each group will share out their tables. What do the tables tell us? Can they make true data statements from the tables?	Bring out vocabulary when appropriate

Next Day Extension:

Using the formula $A = P(1 + r)^t$ and $A = Pe^{rt}$ how can these be used to find the population growth rate. The students should make displays of the data as well as tables.

Data Analysis and Statistical Modeling

Determining Israel's population growth while in Egypt (Genesis 46: 27)

How fast did Israel's population grow while they were in Israel?

1. Genesis 47: 27 tells us that 70 souls of the house of Jacob went into Israel, yet we find in Exodus 1: 7 – 9 tells us that the population of Israel was so great that Pharaoh feared them. How many Israelites do you think would have been in Egypt to make Pharaoh afraid of them?

2. In Genesis 46: 8 – 26 we find the sons of Jacob (called Israel) and the sons and daughters born to them. Make a table of the families and find the **mean** and **median** family sizes. How can this information be used to model the population of Israel during the life of Joseph and his brethren and that generation?

3. Using the mean family size make an estimate of the population size in 100 years.

4. Use the same model to make a prediction of the population in 400 years later than in problem 3.

5. How does the estimate in Problem 4 differ from the population of Israel found in Numbers 1: 19 – 46? Why are they different?

Answer Key for Extensions and Student worksheets

Numbering the Days of a Man, Pt. 1

1. 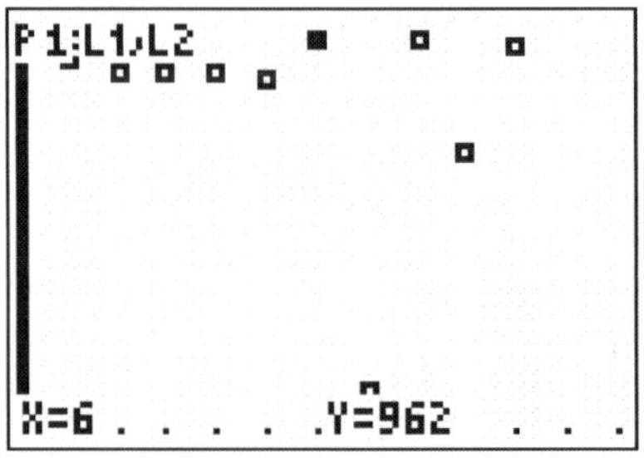

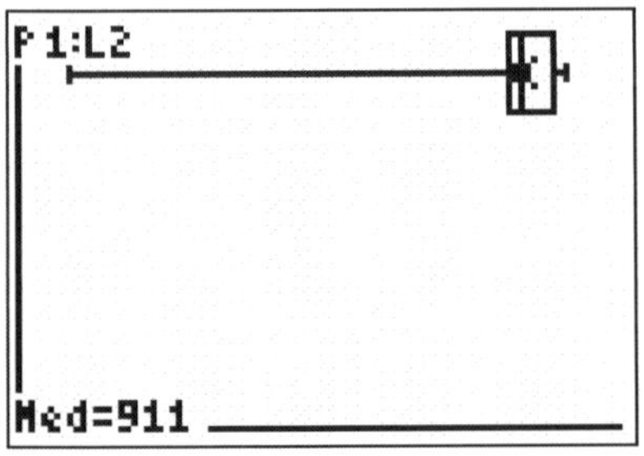

2. The students can use several measures to show distribution of the data set. This set is very compact this can be shown using the IQR which is 55 the median is 911

3. Mean absolute deviation is 107.24 The standard deviation is 166.07

4. Answers may vary

5. The linear regression for the data set is -12.2x + 924.6

Numbering the days of a Man, Pt. 2

1. 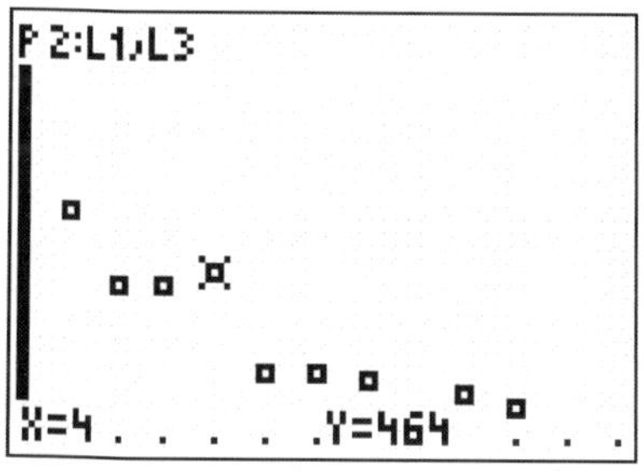

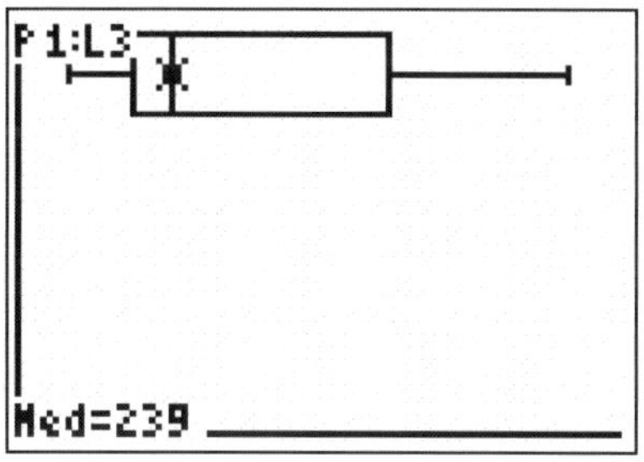

2. A Stem-and-Leaf plot may make the clusters more apparent. This data set has a more pronounced spread. The median is 239. The IQR is 233. There seem to be two distinct clusters and the dividing line seems to be Peleg. After Peleg the lifespan went from 483.75 to 206 a drop of 113%

3. The mean absolute deviation is 133.32 and the standard deviation is 145.19

4. The average lifespan is 317.1 years. The median lifespan is 239

5. An exponential curve fits the Genesis 11 data with an r^2 value of .85 it is obvious from the data set that something happened around the time of Peleg to cause the lifespan to drop significantly. The overall decline of the data is much more pronounced than the data from Genesis chapter 5.

Estimating Pi

1. Answers may vary but the hand breadth must be accounted for when measuring the diameter.

2. Answers may vary. Discuss the challenges of having different standards of a unit measure when it comes to discussing length.

3. Students should discuss the pros and cons of choosing a representative length for the entire class.

4. Answers may vary.

Modeling the Pre-flood population

1. Answers will vary

2. The conditions of the Garden of Eden were perfect there was no death or disease to limit the population. God caused humans and likely animals to have a low birthrate to keep the population from becoming too large.

3. Since there was no death or disease the population would have grown without limits.

4. Once death and disease came into the world, God increased the birthrate so that man would not go extinct.

5. Because of sin, God cursed the ground and caused man to know death.

6. Since the mean lifespan was 857.5 years then they could have reasonably had children for nearly 430 years.

7. Probably about 460 years. (Adam lived for 930 years)

8. Eve could have had around 460/5 or 92 children

The Study of the Cubit and Span

1. Answers will vary.

2. Answers will vary.

3. Answers will vary.

4. Answers will vary.

5. Both measures show deviation but the standard deviation can be moved with very large differences since these differences are squared.

Calculating Usury

1. Answers may vary some credit card interest rates can be as high as 26.99% Department store credit card interest rates can be even higher.

2. $1000(1.06)^2 = \$1123.60$

3. If you are using the $2p = e^{rt}$ then the interest rate can be determined by ——, the solution for 5 years is similar.

4. 25,000 deferred over 4 years is $25000(1+(.06/12))^{48} = \$31,762.23$

5. $P(1+(.065/12))^{36} = \$3,568.51$ $p = \$2937.84$

The Power of Sharing the Gospel

1. Insert calculator image of $y = 2^x$

2. The function is $y = 2^x$

3. Answers will vary.

4. $12(2^x) = 7{,}000{,}000{,}000$ $x \approx 29.11$ days

Determining Israel's Population Growth while in Egypt

1. Egypt was the most powerful country in the world at this time. The Israelites would have had to number in the tens of thousands to have concerned Pharaoh and Egypt.

2. Average family size 5.3, median family size 4.

Jacob	Reuben	Simeon	Levi	Judah	Issachar
14	4	6	3	7	4
Zebulon	Gad	Asher	Joseph	Benjamin	Dan
3	7	9	2	10	1
Naphtali					
4					

3. If each of Jacob's 14 children had 4 children that would be 56, if they each had 4 children that would be 224, then 896, etc until the situation could be modeled as $14(5^x)$. if you figure that the men married and had children at age 25 then you would have roughly 4 generations in 100 years or $14(4^4)$ or 3584

4. In 400 years, roughly 16 generations would have passed this could be modeled by $14(4^{16}) = 6 \text{ X}10^{10}$. This seems like a large number what process is not being considered in our answers? The death rate. The life span in the days of Moses was three score and ten (70) to four score (80).

5. The Bible does not usually record the birth of women. The record in Numbers only lists the men who are old enough to serve in the military (20 years old and older)

Other books in the "Connecting the Bible and Mathematics" Series

Connecting the Bible and Mathematics Through Algebra

Connecting the Bible and Mathematics Through Geometry

Connecting the Bible and Mathematics Through Number Theory

Connecting the Bible and Mathematics Through Measurement

Also from Black Jack Publications

Author and educator Javier Taylor Considers the topic of holiness as it relates the everyday life of a believer.

Made in United States
Troutdale, OR
11/03/2026

56844513R00046